GOOGLE GLASS CAN READ YOUR MIND

DR. JACK M WEDAM

ISBN: 1496173724

ISBN 13: 9781496173720

Library of Congress Control Number: 2014904775

CreateSpace Independent Publishing Platform

North Charleston, South Carolina

CONTENTS

LOOKING INTO YOUR MIND

Until recently, the notion of electronic devices reading your mind was considered futuristic. Brace yourself, because the future is already here.[1] Although the ability for electronic devices to read your mind has been around for several years,[2] this capability was previously confined to laboratories, since the equipment was very expensive and bulky. Furthermore, the clarity with which these machines can read your mind was limited.[3] However, the clarity has improved tremendously recently with new technologies. That is the purpose of this book—to show how much of your privacy is at risk, including your private thoughts, passions, cravings, and other secrets you may want to keep tucked away in your mind.

Much of the development has been driven by safety research that used eye-tracking devices to understand how drivers watch other vehicles and how pilots monitor their instruments. Additionally, marketers study how people respond to visual qualities, such as colors, packaging, and display arrangements. What started out as a way to improve safety and product displays in stores[4] has grown rapidly into something much more sophisticated. Increased computing power facilitated this rapid development, as did the decreased size of equipment, which allowed previously bulky apparatuses to be miniaturized sufficiently to be worn just like ordinary glasses.

The ability to monitor pilots' minds in simulators and to identify cognitive workload requirements was funded by the Office of Naval Research and the Air Force Office of Scientific Research.[5] This resulted in patent protection for methods and apparatuses that can evaluate cognitive activity using cameras capable of recording important eye-tracking data, such as pupil dilation, blinks, eye movements, and fixations.[6] This technology is not limited to military applications; it is now commercially available.[7]

Many companies now offer eye-tracking technology. Consider some of the claims made on various company websites selling their abilities to peer into one's subconscious:

> "Eye tracking can be used as a window into people's subconscious thought processes."[8]

> "EyeTrak uses state-of-the art, infrared eye-tracking technology to accurately measure both conscious and subconscious reactions…"[9]

> "Tobii eye tracking accurately measures both conscious and subconscious reactions to stimuli."[10]

> "By measuring subconscious behaviour, Eye Tracking allows you to monitor whether a message has been seen and remembered."[11]

> "Eye tracking captures shoppers' habitual and subconscious behavior in a natural and unbiased fashion…"[12]

> "Today, however, it is possible for market researchers to get information beyond what consumers say and actually find out how consumers subconsciously feel about what they see…"[13]

You get the idea. However, if you are interested in more examples, searching the Internet and using the terms *eye tracking* and *subconscious* reveals many more claims from a wide variety of companies. After this book is published, you may need to use a search engine other than Google

•   •   •

Google has been very actively pursuing eye-tracking technology.[14] The company was awarded a patent with the innocuous title "Unlocking a screen using eye tracking information."[15] (Hereafter, this title will be shortened to simply "Unlocking a screen.") Perhaps it should rightfully be titled "Unlocking and extracting information from your mind using eye tracking." Uncannily, drawings of this patent with the deceptively innocent title have many similarities to Google Glass. Also, note the bird (item 402) moving from position A to position B, as this will be discussed in more detail below.

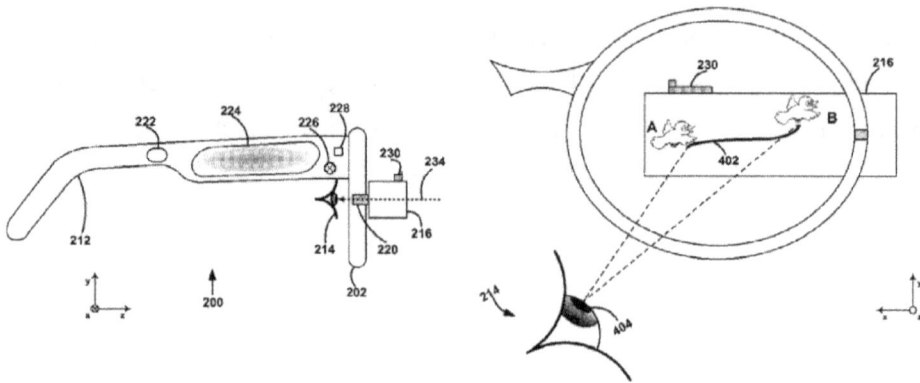

Drawing on page 1 of US Patent 8,235,529[16]

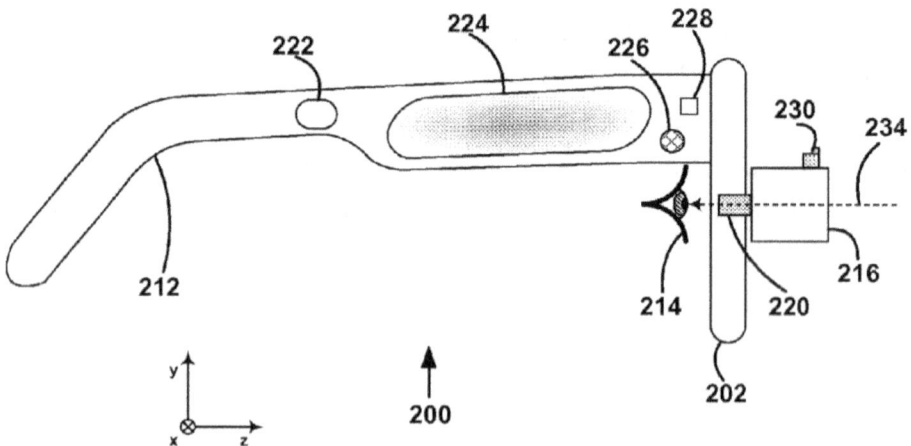

Figure 2B on sheet 2 of 8 of US Patent 8,235,529[17]

A closer look at one of the drawings reveals that this device has a touchpad (item 224 in figure 2B) on the side that is very similar to the one found on Google Glass. Ostensibly, this device is only supposed to unlock a screen using eye-tracking information. The purpose of the touchpad (item 224) relative to unlocking a screen is not explained in this patent and, therefore, it is superfluous to the stated purpose of this patent ("Unlocking a screen"). However, the touchpad is discussed relative to other features, such as a microphone, camera, and speaker,[18] even though there was no explanation in the patent of how a microphone, camera, and speaker would assist in unlocking a screen. Coincidentally or not, Google Glass has a microphone, camera, and speaker.

Google has applied for a variety of patents to protect Google Glass. Samsung does not have a patent titled "Galaxy S5"; neither does Apple have a patent titled "iPad." Google is not required to have a patent titled "Google Glass" in order to protect the various technologies in Google Glass, since it (like others) can essentially mix and match patent protection from various other patents it owns. Predominately, the patents cited in blogs and articles about Google Glass deal with such things as bone-conducting audio, wearable display devices, ornamental design characteristics, transparent displays, and other structural details of Google Glass. Additionally, Google purchased patents for multiuse eyeglasses with input and output interfaces from Motion Research Technologies[19] and head-mounted display technology Foxconn.[20]

The Google patent mentioned above ("Unlocking a screen") is particularly interesting because it has many similarities to Google Glass, yet it has a title that most people would not associate with Google Glass. Furthermore, I could find no evidence that Google is marketing any device with the stated purpose of this very peculiar patent, which is an eyeglass-like device that can unlock a screen. Even though the title is different ("Unlocking a screen" instead of "Google Glass"), is it possible that Google would use this patent to protect some additional features and technology that it may quietly slip into Google Glass? If so, then perhaps we can find some other important clues about the possibilities of reading your mind that Google is, as yet, reticent to publicize.

• • •

There is a multi-billion dollar industry devoted to selling information about consumers to data brokers, marketers and advertisers.

Competition in this industry is fierce, and companies are constantly trying to come up with new ways to mine such information.[21] Is Google seeking ways to collect[22] very personal information on you in order to sell that information in this competitive market? If Google were able to read people's minds and sell such very private and personal information gathered in this way, it would have a huge advantage over its competitors. As just one example of such possibilities, one company has already developed an app that it claims can detect your emotions using Google Glass and then relay that information back to retailers.[23]

Intent is difficult to prove, but perhaps many descriptions and claims in the eye-tracking device patent were serendipitous. Perhaps Google developed and patented this technology for uses that were completely innocent, and Google really didn't grasp that Google Glass would enable computers to read your mind. However, if the personnel at Google were not already aware of how this technology could be used to read your mind, then they certainly will be after they read this book. Likewise, if hackers were not aware of it before the publication of this book, they soon would have figured it out on their own. Then we would be vulnerable to unscrupulous hackers who could read your mind in real time by hacking into Google Glass. What might they do with the secrets buried deep in your subconscious?

· · ·

Google Glass was introduced in the spring of 2013 with much publicity, well-designed websites that featured superb photography and videos, and pictures of glamorous women wearing the eyewear. These glasses looked exciting, as if they were the important next step into the future. If you paid any attention to the rollout, you probably saw what Google wanted you to see. However, sometimes what's deliberately not mentioned is the most important question: Is there more to this story?

A closer examination of all the facts may surprise you.

An important legal point must be stated before going on. For legal reasons, I am not alleging that Google is currently using eye-tracking technology to read your mind. Instead, this book is meant to provide a technological assessment of how recently patented eye-tracking technology could be used in hypothetical situations. However, you should be aware that US Patent 8,235,529 ("Unlocking a screen") awarded to Google protects technology that could enable computers to read your mind in some hypothetical situations. The technology could do so by closely monitoring what you're gazing upon (gaze axis) and the jerky movements of your eyes, which reveals what your subconscious is watching. Although there was no public mention of the fact that this new technology can read your mind, Google could use it anytime it chooses, and you would not be able to detect it. Even if Google chooses not to use this technology, it's just a matter of time until hackers will, as it will likely be too tantalizing for some of them to resist.

Two researchers recently developed a proof-of-concept spyware app to demonstrate that Google Glass can secretly take photos of whatever a Google Glass wearer is gazing upon without the wearer's knowledge. Fortunately, these were not nefarious hackers. Instead, they were respected researchers who publicly revealed their app as a warning to the public.[24] It is nearly impossible to protect computer systems everywhere from hackers.[25] Reuters has reported on numerous large-scale cyber-attacks in recent years. On February 25, 2014, Reuters reported that 360 million newly stolen credentials were on the black market as a result of a sophisticated cyber-attack.[26] It is bad enough when we learn that our personal financial information was hacked from corporations or governments[27] that promised us our information was secure. Now we learn that Google Glass has also already been hacked.[28] Hackers may be able to gain many powers that Google never intended anyone to have.[29] How will you feel when you find out that your mind has been hacked? Worse yet, hackers will not only be able to extract your conscious thoughts out of your mind, but also your subconscious thoughts as well.

DON'T BE JERKED AROUND

The unwitting may be fooled by this new technology, but you need not be if you pay attention for a few more pages. The use of the word "jerked" in this context is appropriate, since measuring tiny, jerky eye movements is how a computer can read your mind. The French word *saccadé* is used to describe an abrupt pull on the reins by a horse rider. In the 1880s, a French ophthalmologist, Louis Émile Javal, used the term to describe the jerky motions of the eye.[30] For many years, however, the purpose and function of these jerky motions remained a mystery.

Research into saccades in recent years has accelerated tremendously.[31] Some experts consider microsaccades, saccades, and fixational saccades to have somewhat different functions but many similarities.[32] Most experts consider them to be related.[33] If you would like to learn more, read the excellent article, "The impact of microsaccades on vision: towards a unified theory of saccadic function," in the February 2013 edition of *Nature*.[34] Fortunately, you don't have to understand all of the nuances in order to understand how these jerky movements can reveal much about your thoughts. For simplicity, in this text, "saccade" will be used for all three terms.

Science has recently unraveled the mystery of how short, jerky eye movements that increase clarity[35] can also cause some optical illusions.[36] Rather than explaining the phenomenon from a scientific point of view, perhaps it would be better to look at it in the context of common experiences. Saccades are responsible for the optical illusion by which airplane propellers or other rotating objects occasionally appear to spin backward.[37] The short, jerky movements keep vision sharp. If the eyes didn't make small movements, objects in your view would fade.[38] When light enters the eye, it hits the retina, which contains rods and cones. When light strikes the rods and cones, it causes the discharge of electrical impulses that travel through the optic nerve and then on to the brain. The brain integrates all of these electrical pulses into what the brain perceives as a visual image. The problem is that once the rods and cones in the retina discharge, they need a little time to recharge. Therefore, the eye moves slightly so that the light will fall on fresh rods and cones next to those that have recently discharged. This gives the previously discharged rods and cones time to recharge. Amphibians and reptiles do not have this feature and so can only see things when they move. This is only one function of saccades.

Their second function of these rapid eye movements is more relevant to the topic of this book. Saccades are an integral part of peripheral vision. Have you ever been looking attentively at something directly in front of you and then suddenly catch a glimpse of something out of the corner of your eye? This phenomenon is made possible by saccades.

Attention is a selective process. The conscious mind simply can't process all of the visual information that the eyes provide. Many scientist estimate that that 95 percent of all cognition occurs in the subconscious mind.[39] To handle this massive amount of visual information, the mind has developed the neurological equivalent of a triaging system to sort out and prioritize stimuli.

When referring to visual attention, "covert" does not mean "sneaky"—it merely means that you're not consciously aware of what's going on. However, your subconscious is very aware of what's in your peripheral vision, even though you're not gazing directly at any object found there. Thus, the conscious mind can stay focused on the small area that's of immediate interest to the viewer (overt attention), while things in the peripheral vision are handled by the subconscious (covert attention).[40] If you want to know more, a review article, "Visual attention: The past 25 years" by Marisa Carrasco in *Vision Research* may be of interest to you.

Analyzing your covert attention requires a device that can precisely measure the movements of your saccades. Moreover, the claims in the summary portion of Google's patent 8,235,529 ("Unlocking a screen") indicate that's what's happening in the process of using the eye-tracking information. However, the process used to unlock the screen is merely the calibration phase of a more involved process that bypasses the cumbersome manual process previously required by other eye-tracking devices. A manual calibration process would provide a warning to users that the device was about to start reading their minds. This patented technology automates the previous cumbersome manual process and makes this step invisible to the user.

WHY DID GOOGLE STUFF
THIS PATENT LIKE A TURKEY?

The US Patent and Trademark Office awarded Google US Patent 8,235,529 ("Unlocking a screen") on August 7, 2012, which was just a few months before the big, splashy public rollout of Google Glass. Patents are often very long, excruciatingly technical, and difficult-to-read documents. Typically only patent attorneys, who are paid well for their time, bother to take on this feat. However, if you know what to look for and where to look, patents can be a proverbial treasure chest of information. For legal purposes, I must disclose that I'm not a patent attorney. However, having gone through the patenting process, and having been awarded a patent myself,[41] I can show you some of the salient features of patents and point you to where you may find some interesting information. You're welcome to read all of the Google patents from beginning to end if you have the time and tenacity, or I can point out the highlights so that you don't get overwhelmed and give up before you get to the pertinent information.

This patent for unlocking a screen uses the term "eye tracking" eighty-eight times and some form of the word "calibrate" another thirty-three times. Of course, there is nothing nefarious about eye-tracking devices in and of themselves. These devices have been around for years, and they have been used for many different purposes, as mentioned above.

However, one of the problems with using eye-tracking devices is manual calibration.[42] Before eye-tracking devices can be used, they must first be calibrated so that the computer can be sure that it is accurately tracking what the person is looking at. This can be accomplished by instructing a person to look directly at a spot on a screen and then at another spot in a different location, followed by more spots in more locations until the calibration is complete. Alternatively, an individual can be instructed to look sequentially at various letters on a wall chart.

There are two drawbacks to this manual method: it takes time, and it is a sure tipoff to users that their eyes are being calibrated. The latter is not a problem if the person has agreed to this calibration. In fact, when using a manual method, it's almost impossible to calibrate a person's eyes without consent and cooperation. However, the methodology described in US Patent 8,235,529 allows an eye-tracking device to be calibrated without the subject's knowledge. The glasses unlock the screen once your eyes are calibrated either by having you watch a bird move across the display from point A to point B, as shown in item 402 in the drawing mentioned above and figure 3 of the patent (below), or by reading text, as shown in figures 5 and 6 of the patent (below). Both the bird (item 402 in the patent) moving along a path and your eye movement associated with reading text will be discussed in more detail later. Feel free to refer back to these figures.

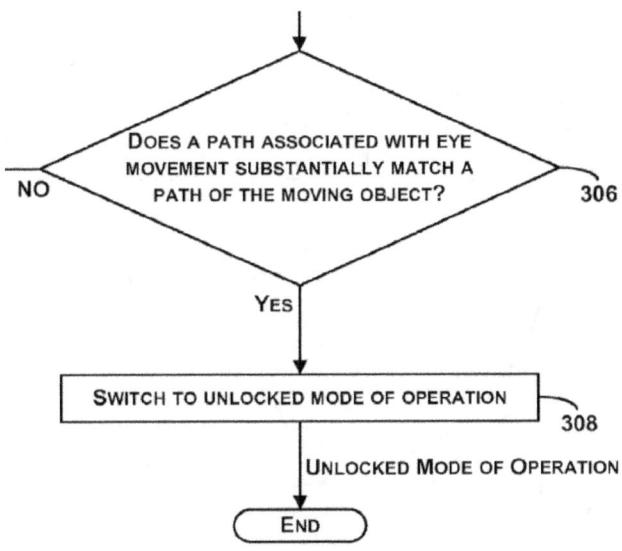

Extracted from figure 3, sheet three of eight, US Patent 8,235,529[43]

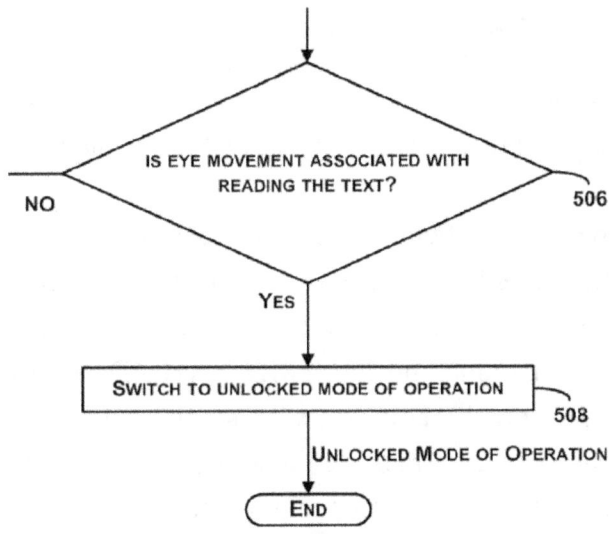

Extracted from figure 5, sheet five of eight, US Patent 8,235,529[44]

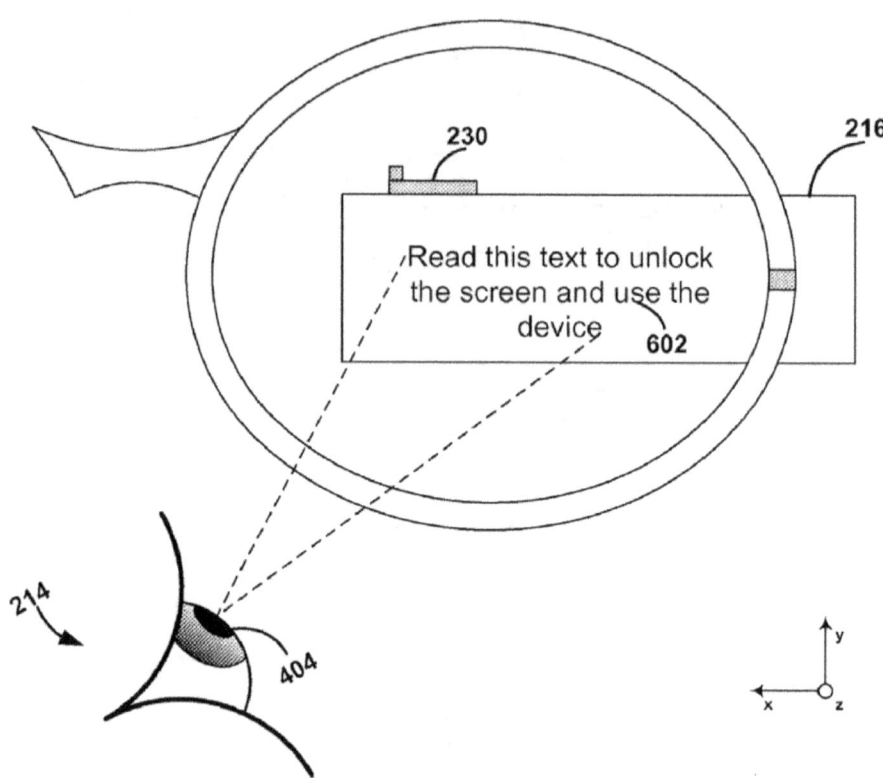

Extracted from figure 6, sheet five of eight, US Patent 8,235,529[45]

Google does not explicitly state in the claims portion of its patent that the system calibrates the eye of a subject. However, elsewhere in this patent, Google states, "calibration of the system may be integral to or byproduct of unlocking the HMD."[46] (HMD is an acronym for "head-mounted display.") This is a very significant issue because one of the problems with eye-tracking devices previously had been time-consuming manual calibration that requires the cooperation (and, therefore, the willing consent) of the people using the devices. However, what's described in claims one through eight indicates how this device can be calibrated. Ostensibly, once this process is complete, the screen is unlocked.

To put this into perspective, Google has patented a process to calibrate your eye every time you read text displayed on its screen. You may think you are asking for information and seeing the information quickly pop up on the display, but the device can use the opportunity to calibrate your eye. Therefore, this process allows Google to bypass the manual calibration process and do so without your awareness or consent.[47] Keep in mind that this patent protection can be transferred to any other Google products, such as Google Glass. Furthermore, this capability would not be limited to Google Glass. Google could use this eye-calibrating technology in any software or operating system it has. Any smartphone that has eye-tracking capability and uses Google's Android operating system could potentially be turned into a mind reader. *Instead of just reading your smartphone, your smartphone could be reading you,* [48] *if it has eye-tracking capabilities.*

It is important to understand the difference between *eye-tracking* and *head-tracking* capability. The new Amazon smartphone will likely have *head-tracking* capability.[49] I could not find any literature that suggests head tracking is used for mind reading. If some corporation wants to try, the mind reading clarity would likely be very low and be limited to whether or not you are paying attention. Head tracking capability would be useful for stopping videos when you were distracted, such as turning your head to look around. Therefore, you can be at ease with Amazon's head-tracking capability versus Google's eye-tracking capability.

•   •   •

Patents' claims are a mechanism that provides legal protection to the inventors, "as it is the claims that define the scope of the protection afforded by the patent and which questions of infringement are judged by the courts."[50] If Google just wanted to provide a more convenient way for you to unlock your computing system after a period of inactivity,[51] it could have stopped at claim eight, as that is where it states that the screen is unlocked.

Let's do a little math: this patent has twenty-seven claims, but less than a third of the claims are devoted to explaining how the glasses get to the point of unlocking the screen. So why the extra claims describing other eye-tracking functions *after* the apparatus has unlocked the screen? After all, the title of the patent is "Unlocking a screen using eye tracking information."

Perhaps we can find a clue as to what else these glasses can do by looking at the wording in the claims following the unlocking of the screen. At the end of claim eight, Google states that after switching from locked to unlocked mode, the glasses can track your eyes. This is not my interpretation but Google's own words.

Patents don't have page numbers; they have columns and line numbers instead. Citing claim eight in the patent, this device can "switch a mode of operation of the computing system from the locked mode of operation to the unlocked mode of operation; and *enabling the eye tracking system.*"[52] (Emphasis added to column 16, line 35–39, US Patent 8,235,529.)

This patent is loaded with many other claims that protect Google's eye-tracking technology, including various aspects of tracking, communicating, and recording information about your eye movements. Consider some other claims in this patent (after the device has unlocked a screen in claim eight) and remember that the stated purpose is merely to unlock a screen:

- "*eye tracking system* comprises at least one sensor configured to monitor *eye movement.*" (Ibid. column 16, lines 61–63, emphasis added.)

- "the computing device includes an *eye tracking system*; receiving information associated with *eye movement* of a user of the computing device from the *eye tracking system*..." (Ibid. column 17, lines 6–10, emphasis added.)

- "an *eyetracking system* in *communication* with the wearable computer, wherein the *eye tracking system* is configured to track *eye movement* of a user of the wearable computer...[and] receive information associated with the *eye movement* from the *eye tracking system*..." (Ibid. column 18, lines 1–12, emphasis added.)

- "enable the *eye tracking system* so as to receive the information associated with *eye movement* from the *eye tracking system*..." (Ibid. column 18, lines 35–38, emphasis added.)

- "the wearable computing system includes an *eye tracking system*; receiving information associated with *eye movement* of a user of the HMD from the *eye tracking system*..." (Ibid. column 18, lines 49–52, emphasis added.)

• • •

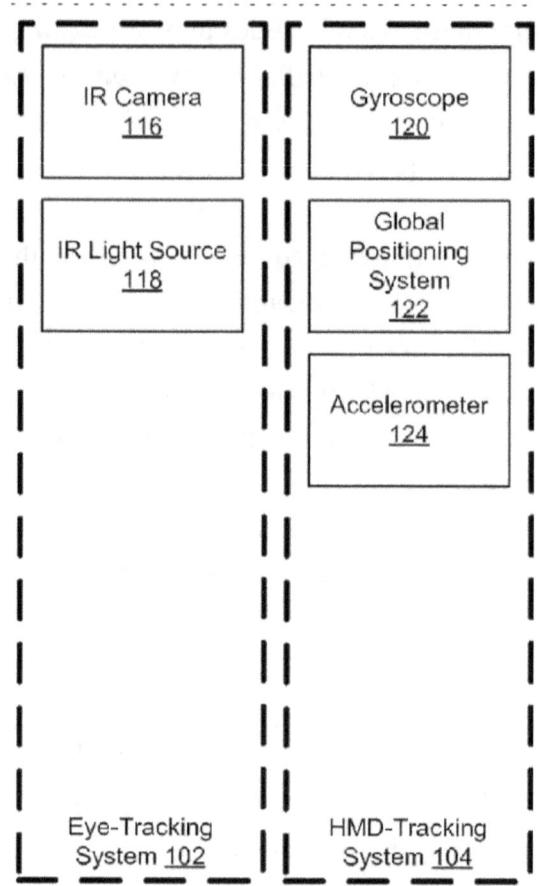

Extracted from figure 1 of US Patent 8,235,529[53]

According to figure 1 on sheet one of eight of US Patent 8,235,529, this eyewear, which ostensibly merely unlocks screens, includes many other things: gyroscopes, a global positioning system, and accelerometers. It's hard to imagine how the combination of gyroscopes,[54] GPS, and accelerometers would be needed or even helpful for unlocking a computer screen. The eyewear described in US Patent 8,235,529 is not being marketed for the purpose stated. Although, rather coincidently, this patent describes many features[55] found in Google Glass, such as GPS, gyros, and accelerometers.[56] (As of this time, Google Glass does not have a *built-in* GPS. However, GPS capability is available by Bluetooth when connected to other devices.[57]) It appears that Google used this so-called screen-unlocking patent to protect many of the features in Google Glass (GPS, gyros, and accelerometers). These things are superfluous to the stated purpose of this patent—unlocking a screen—unless perhaps Google believes you may do such crazy things as beat your head against a wall (accelerometers), then tilt your head back and forth (gyros) as you run around the neighborhood (GPS) in order to unlock your screen. Nevertheless, remember the gyros and accelerometers, since they will be discussed again with a possible explanation why they were included in this patent.

•   •   •

This patent to unlock a screen using eye-tracking information was assigned to Google with Hayes Solos Raffle listed as one of the three inventors. A LinkedIn page states, "Hayes Raffle leads the interaction research team conducting advanced development for Google Glass."[58] The second inventor listed on the patent is Adrian Wong. A LinkedIn page for Adrian Wong states, "Google Glass—Hardware Engineering Manager (EE and Systems Engineering)."[59] The third inventor listed on the patent is Ryan Geiss. A website that appears to be his states, "I currently work at [a] place known as the Google, on a team which might or might not exist, on Google Glass and the other things."[60] The nexus of Google Glass, these three inventors, and a patent involving eye-tracking technology does not prove that Google Glass will use eye-tracking technology. Nevertheless, the nexus is worth noting. These postings may disappear after this book is published. If these three inventors are willing to stand up to public scrutiny, they will leave these postings on the Internet. If they remove these postings, it will raise questions why the postings were removed.

There are so many similarities between Google Glass and the patent to unlock a screen that perhaps we should examine the other details of this patent in order to get clues about other things that Google may have up its sleeve for Google Glass.

"SKILLED IN THE ART"

In patent speak, the term "skilled in the art"[61] is used to describe those people who have more knowledge about certain technical areas than a layperson but not as much as a genius.[62] The meaning here is that it's not necessary to go into detail explaining things if it would be obvious to those already "skilled in the art." Here is the commonsensical interpretation: when the phrase "those skilled in the art would understand" is used, what follows is often summarized in a general way rather than a specific or detailed manner. This provides a great deal of legal wiggle room. If the patent is contested in the future, the lawyers defending the patent can point out that something was mentioned even if the text was vague and lacking in clarity. Perhaps this means they just didn't want to disclose many details because, ostensibly, it's obvious to anyone familiar with the technology or process described in the patent. If so, using the phrase "those skilled in the art would understand" provides a very clever way to build in some protection without revealing too much detail. So perhaps more closely examining this patent to look for the phrase "skilled in the art" will provide us with more clues as to what Google may be trying to use this for in the future.

Consider three places in US Patent 8,235,529 ("Unlocking a screen") where the term "skilled in the art" is used:

1. "Those skilled in the art would understand that *other user input device,* user output devices, wireless communication devices, *sensors, and cameras* may be reasonable included in such a wearable computing system." (Ibid. column 7, lines 57–61, emphasis added.)

2. "As such, those skilled in the art will appreciate that other arrangements and *other elements…*can be used instead." (Ibid. column 15, lines 29–32, emphasis added.)

3. "Other aspects and embodiments will be apparent to those skilled in the art. The various aspects and embodiments disclosed herein are for purposes of illustration and are *not intended to be limiting,* with the true scope being indicated by the following claims, along with the full scope of equivalents to which such claims are entitled. It is also to be understood that the terminology used herein is for the purpose of describing particular embodiments only, and is *not intended to be limiting."* (Ibid. column 15, lines 39–47, emphasis added.)

These three references to "those skilled in the art" open very large legal doors, essentially meaning that Google can add almost anything it wishes to this device. For example, it could add sensors in order to measure pupil response and detect eye blink. Your pupils' response to images can reveal an amazing amount of information about your thoughts and preferences.[63] Your pupils and blink rates can reveal much about what you are trying to conceal.[64] Moreover, the way Google has worded this patent, it can include sensors that can detect your blink rate and changes in your pupils. These technologies are wonderful if they protect us from criminals or terrorists; however, it would be very different if these technologies were used against us. Would you want retailers to know that you really are interested in buying something they are selling even though you are trying to resist and conceal that information from them?

Consider some interesting possibilities: by putting on Google Glass, you could be stepping right into a scene from *Person of Interest or Minority Report*. However, it would not be science fiction, and you would probably not be as lucky as the main characters. Although you may not have murder on your mind, someone you do not know remotely monitoring your mind could learn many things about you from both your conscious and subconscious mind. This could include insights into your temptations, cravings, and strong urges. Retailers would likely pay a premium for such information. Worse yet, what would nefarious hackers do if they were able to pry into your mind?

OVERT AND COVERT ATTENTION

O vert attention can be detected by monitoring what you're looking at—in technical terms, monitoring your gaze axis. Head-mounted displays fitted to consumers have been used by marketers to learn how consumers look at individual products, product packaging, or products in displays, such as store shelves. More recently, eye-tracking devices have been used to monitor how people look at web pages. By using these tools, researchers can refine and improve the visual image of web pages, products, and packaging as well as determine how their product stands out on the shelf compared to a competitor's product. A computer records exactly where the eyes are looking, and these records are then turned into what are known as eye-tracking heat maps, eye heat maps, or just heat maps.[65] Heat maps use various colors overlaid on images in concentric zones to indicate areas of increased attention by noting longer duration of gaze upon a particular spot. These are somewhat similar to tornado or hurricane maps indicating locations with a high probability of being affected with concentric zones of color or shading. Whereas the weather maps indicate where severe weather may move, eye heat maps indicate exactly where your gaze remains upon an image. Longer durations of gazing at one spot result in brighter colors overlaid onto the image.

Google's patent for unlocking a screen with eye-tracking information uses the term "gaze" thirty-nine times and "gaze axis" twenty-one times, which is a huge number for a patent that ostensibly just unlocks your screen. This suggests that the technology reflected in this patent is very capable of detecting what you're looking at by analyzing your gaze axis. Gaze axis correlates to overt attention, since what you're gazing at is usually the object of your conscious mind.

Manually detecting overt attention is not new. Peering into people's eyes to get a hint of what's in their minds has been common practice for thousands of years. However, computers can now very accurately monitor overt attention.

Gaze axis, therefore, provides detailed information about not only what one is looking at, but also how intently one is looking at various *parts* of an object. This provides a huge insight into one's conscious mind. You're encouraged to look at some heat maps to see the level of detail they provide[66] as a caution of how exactly Google Glass could potentially record your visual activity.

Therefore, with the technology protected in this patent (which ostensibly just unlocks your screen), Google Glass can assist a computer in constructing a heat map showing precisely where your eyes are looking and how intent your gaze is. The computer can do all of this without you even being aware.

To put this into perspective, if you are wearing Google Glass, someone (or a computer) could remotely monitor who you are looking at, what part of the person's body you are looking at, and how intensively you are doing so. This sort of information would likely be very valuable to marketers. Worse yet, it could cause you many problems if this sort of information was nefariously obtained by hackers.

What about covert attention? Some people have been able to catch a glimpse of another's covert attention by closely watching their shifty eyes. However, this type of monitoring was much less reliable than just noting what the person's eyes were looking at directly. The shifty eyes were just too quick for humans to follow accurately. However, that's all changed recently with this new technology.

•   •   •

A musical metronome is a device that musicians use to keep a steady tempo. They can be adjusted for various beats per minute. Using a musical metronome set to sixty beats per minute or one beat per second, an eighth note is five hundred milliseconds, or half of a second, since a second can be divided into one thousand intervals, each being one millisecond long. A sixteenth note would be 250 milliseconds, which is represented as 0.250 seconds, 250 msec, or just 250 ms. Many people can perceive a thirty-second note, which lasts 125 milliseconds. Although they are of very short duration, some can even perceive sixty-fourth notes, which only last 62.5 milliseconds[67] when the musical metronome is set to sixty beats per minute. Please feel free to refer back to this section, as you will see these durations again soon.

So what was the purpose of that short musical lesson? It provides a way to take you from something that you're familiar with to something that you probably intuitively understand but that uses different terms to describe the same phenomenon. Since patents are stuffed with legalese and are often very difficult to read, you need to be acquainted with such terms before delving into the patent itself. Moreover, the length of musical notes is something with which most people are familiar. With that as our basis, let's examine the phenomenon of saccades more closely.

There is simply too much in your field of view for your conscious mind to make sense of it all, especially since your conscious mind handles only about 5 percent of the cognitive workload of your brain while 95 percent is handled by your subconscious. In a simplified explanation, your mind triages what's most important to you, and what your mind triages to your subconscious becomes your covert attention.

Perhaps this analogy will make it easier to understand: think of your subconscious as a personal assistant who will monitor things going on about you but will not interrupt you unless it's something very important. For example, if you were watching a baseball game and focusing intensely on the pitcher, your conscious mind may not notice your friend waving to get your attention at the end of the row. However, your subconscious is aware that someone is waving and will bring the action to your conscious attention. Likewise, if you're riding a bicycle and trying to catch up with the rest of your group, you may not see a car coming at you from the side. However, your subconscious would, and it would inform your conscious mind to protect you from a potentially dangerous situation.

Your conscious mind stays focused on a very small part of your visual field. However, your subconscious is very actively monitoring the remainder of the visual field—what is commonly known as your peripheral vision. The problem is that your eyes can only look at one spot at a time. Therefore, your conscious mind and your subconscious mind both utilize your eyes on a shared basis. Your eyes quickly wander in order to monitor the periphery; most people are not even aware of their exploratory saccades.[68] However, you can consciously override[69] this automatic subconscious exploration of your peripheral vision.

In the case of saccades, covert attention means, "not openly acknowledged or displayed."[70] In this case, people are not consciously aware of the small movements of their own eyes or of the other details within their fields of view beyond what they're already focused on. The subconscious exploration of the peripheral vision is largely involuntary.[71] Covert attention is also called reflexive attention[72] as the attention shifts reflexively to different areas of the peripheral vision as new objects enter the visual field or move. These subconscious, exploratory saccades can be very accurately measured along an index of covert attention.[73] This can be accomplished by computers, since algorithms have been developed that can accurately detect saccades.[74]

The rate of saccades changes as your subconscious becomes more interested in things in your peripheral vision. As something new comes into your peripheral vision, the rate of saccades decreases for a short while and then increases above normal rate once again.[75] The direction of saccades is an indicator of either subconscious or covert attention.[76] Covert attention shifts can be identified by a change in rate of the saccades as compared to a baseline for a particular individual.[77] Saccades can also reveal orientation of covert attention.[78] Therefore, saccades can reveal what has your attention, and they can reveal your subconscious thoughts,[79] including what you're trying to resist looking at.[80] For example, if you're trying to lose weight and you spot your favorite dessert sitting on the table in your peripheral vision, your subconscious may try to resist looking at the dessert just as if you told your personal assistant, "Don't look over there!" If you're a shopaholic and trying to gain control over your spending, your saccades can reveal that temptation to marketers. If you're prone to drinking too much alcohol but you're trying to control your alcohol consumption, your saccades could reveal not only your temptation to alcohol but your favorite drink to marketers. If you're addicted to pornography and feeling guilty about it, your saccades might indicate that as well. Therefore, Google's new technology could be used to reveal your temptations and vulnerabilities, which marketers could then exploit.

Does the technology reflected in Google's patent 8,235,529 have the ability to detect saccades? Indeed it does.

The word "saccade" is used nine times in this patent, "ms" (milliseconds) is used three times, and "milliseconds" is used once. Here are a few examples of how this patent uses the term "saccades":

- "Tracking a slowly moving object may reduce a probability of eye blinks, or rapid eye movements (i.e., *saccades*) disrupting the eye-tracking system." (Ibid. column 3, lines 50–53, emphasis added.)
- Recalling the brief explanation above on duration of musical notes, a sixteenth note was 250 milliseconds. This patent describes the ability to measure the speed of saccades very accurately. "Once *saccades* start, fast eye movement may not be altered or stopped. *Saccades* may take 200 milliseconds (ms) to initiate, and then may last from 20–200 ms, depending on amplitude of the *saccades* (e.g., 20–30 ms is typical in language reading)" (Ibid. column 9, lines 31–35, emphasis added.)
- *"Saccades* may disturb or hinder an ability of the eye-tracking system…to track eye movement. To prevent such disturbance to the *eye tracking system*,…the processor may generate the display of the moving object such that the speed of the moving object may be below a predetermined threshold." (Ibid. column 9 lines 35–40, emphasis added. Also, the number "230," which correlates to an item on drawings, was deleted for simplicity.)
- The virtual bird (item 402 in US Patent 8,235,529) that moves across the field of view from point A to point B at a very specific rate is just perfect for detecting the direction of one's eye movement and rate. "The processor may display the bird moving at a speed that may match an ability of a human eye to follow the moving object without *saccades*." (Ibid. column 9, lines 12–15, emphasis added.)
- "If the speed exceeds the predetermined threshold speed, *saccades* may be stimulated." (Ibid. column 9, lines 40–43, emphasis added.)

Therefore, the technology protected in this patent can differentiate between normal eye movements and saccades. It can even tell the threshold for each individual to distinguish saccades from nonsaccade rapid movements of the eye.

This opens the door technologically for measuring saccades and, therefore, to your subconscious. This recently patented technology provides a path into your mind where computers could bountifully harvest your private thoughts, as well as your fears, subconscious passions, and cravings. Whether or not that's what Google intended, we'll probably not know in the near future. However, we may find out after it is too late.

WHAT THE FUTURE MAY HOLD

With a few modifications, these technologies would not be limited to Google Glass or smartphones. This technology could be scaled up and used in any public place that has video display. Remember that computers are always getting smaller, and sensors are always getting better. Therefore, any video in any mall, train station, airport, or other public place could be utilized to read your mind soon if it can't already. Consider that any moving objects, such as birds moving across a large public display or text displayed, may be used to calibrate your eye. Couple this with facial recognition technology, and your mind could be read in any public place with a video display.

Instead of large public displays, consider the possibilities of miniaturization. Do you recall the accelerometer and gyroscope that were inexplicably inserted into the patent to unlock a screen with eye tracking information? These would be useful in measuring saccades if Google reduces Google Glass small enough to insert into contact lens. A miniaturized camera that could move with your eye would be able to record the object of your covert attention. Although Google Glass has a camera, it is looking forward similarly to your overt attention and may miss objects in your peripheral vision. A camera that moves with your eye would be much better suited for recording your covert attention if objects were near or outside the "peripheral vision" of the camera on the regular Google Glass. Apparently, this has not escaped Google's attention, as Google recently announced their secretive Google X lab has miniaturized Google Glass (including the camera) so that it could fit on a contact lens.[81]

• • •

The assessments of technological capabilities contained in this book may elicit howls of denial from Google. Therefore, I have provided extensive documentation with references to many scientific papers, line-by-line references to Google's patent, as well as several examples of other companies that are offering electronic mind-reading services. You are encouraged to read and learn even more so you do not have to rely solely on carefully parsed statements from Google. After you learn more, you should draw your own conclusions and make your own assumptions.

If you do not have the time to dig deeper and learn more, you should, at a minimum, be very discerning of any carefully parsed statements from Google that may be designed to protect its market share and corporate image in a fiercely competitive market instead of revealing the whole truth. Consider the company's behavior up to this point in time. Consider that it prides itself on being at the forefront of innovation with its secretive Google X Lab.[82] Consider how it has already encroached upon your privacy. Consider how competitive the personal information business is now and will be in the future. Furthermore, consider how profitable it would be if Google decided to use these technologies to read your mind and sell your private thoughts to marketers. Also consider, however, if Google does not utilize technologies that could read your mind, hackers likely will soon if they haven't already.

• • •

NOTES

(Endnotes)

[1] Richard Gray, "Computers that read minds are being developed by Intel," *The Telegraph*, August 22, 2010. Accessed March 3, 2014, http://www.telegraph.co.uk/technology/news/7957664/Computers-that-read-minds-are-being-developed-by-Intel.html.

Nic Halverson, "Future Eye-tracking Systems Will Read Your Mind," *Discovery*, March 28, 2012. Accessed May 3, 2013, http://news.discovery.com/tech/apps/future-eye-tracking-120328.htm.

[2] Daniel Bor, "The Mechanics of Mind Reading," *Scientific American*, July/August 2010. Accessed April 10, 2014, http://www.scientificamerican.com/article/the-mechanics-of-mind-reading. (Payment required.)

[3] Weintraub, Karen. "Scientists: Mind reading might not be far off" *Detroit Free Press*, April 22, 2014. Accessed April 23, 2014. www.freep.com/article/20140422/FEATURES08/304220051/Scientists-explore-possibilities-of-mind-reading

[4] Kris Van Cleave, "Eye-tracking technology helps marketers and medical professionals alike," *WJLA*, May 7, 2012. Accessed May 6, 2013, http://www.wjla.com/articles/2012/05/eye-tracking-technology-helps-marketers-and-medical-professionals-alike-75702.html.

[5] Grant No. N00014-93-1-0525 from the Office of Naval Research and Grant No. F49620-97-1-0353 from the Air Force Office of Scientific Research. Sandra P. Marshall, US Patent 6,090,051 "Statement of Government Rights: Method and apparatus for eye tracking and monitoring pupil dilation to evaluate cognitive activity," filed March 3, 1999 and issued on July 18, 2000. Accessed April 9, 2014, http://patft.uspto.gov/netacgi/nph-Parser?Sect1=PTO1&Sect2=HITOFF&d=PALL&p=1&u=%2Fnetahtml%2FPTO%2Fsrchnum.htm&r=1&f=G&l=50&s1=6,090,051.PN.&OS=PN/6,090,051&RS=PN/6,090,051.

[6] Sandra P. Marshall, US Patent 6,090,051, "Method and apparatus for eye tracking and monitoring pupil dilation to evaluate cognitive activity" filed March 3, 1999 and issued on July 18, 2000. Accessed April 9, 2014, http://patft.uspto.gov/netacgi/nph-Parser?Sect1=PTO1&Sect2=HITOFF&d=PALL&p=1&u=%2Fnetahtml%2FPTO%2Fsrchnum.htm&r=1&f=G&l=50&s1=6,090,051.PN.&OS=PN/6,090,051&RS=PN/6,090,051.

[7] "SOFTWARE: Cognitive Workload," EyeTracking, Inc. Accessed March 25, 2014, http://www.eyetracking.com/Software/Cognitive-Workload.

[8] "Surprise! Eye Tracking Shows Men Look at Women and Men," *Eye Tracking Update*, August 21, 2010. Accessed May 8, 2013, http://eyetrackingupdate.com/2010/08/21/surprise-eye-tracking-shows-men-women-men.

[9] "Eyetrak™ Visibility Research," Visuality Group. Accessed May 8, 2013, http://www.visuality-group.co.uk/shopper-behaviour-and-research/eyetrak.

[10] "Eye Tracking," Tobii. Accessed March 4, 2014, http://www.mydelphi.eu/eye-tracking.html.

[11] "Eye Tracking: What catches your consumers' eyes?" BDRC Continental. Accessed March 4, 2014, http://www.bdrc-continental.com/technology-analytics/eye-tracking.

[12] "Eye Tracking Demo at Shopper Marketing Live, May 19–20," Tobii. Accessed May 8, 2013, http://www.objectivedigital.com/ot_may2011.htm.

[13] "Eye Tracking: A New Method for Marketers to Identify How You Really Feel About the Media You See," Eye Tracking Update, October 15, 2009. Accessed May 8, 2013, http://eyetrackingupdate.com/2009/10/15/eye-tracking-a-new-method-for-marketers-to-identify-how-you-really-feel-about-the-media-you-see.

[14] Chris Wood, "Google eye tracking unlock patent revealed," *Gizmag*, August 8, 2012. Accessed January 5, 2013, http://www.gizmag.com/google-eye-track-unlock-patent/23637.

[15] Hayes Solos Raffle, Adrian Wong, and Ryan Geiss, US Patent 8,235,529, "Unlocking a screen with eye tracking information" filed November 30, 2011 and issued on August 7, 2012. Accessed March 3, 2014, http://patft.uspto.gov/netacgi/nph-Parser?Sect1=PTO1&Sect2=HITOFF&d=PALL&p=1&u=%2Fnetahtml%2FPTO%2Fsrchnum.htm&r=1&f=G&l=50&s1=8,235,529.PN.&OS=PN/8,235,529&RS=PN/8,235,529.

[16] Hayes Solos Raffle, Adrian Wong, and Ryan Geiss, US Patent 8,235,529, "Unlocking a screen with eye tracking information" filed November 30, 2011 and issued on August 7, 2012. Accessed April 9, 2014, http://pdfpiw.uspto.gov/.piw?PageNum=0&docid=08235529&IDKey=DB61E66D7179%0D%0A&HomeUrl=http%3A%2F%2Fpatft.uspto.gov%2Fnetahtml%2FPTO%2Fpatimg.htm

[17]   Ibid.

[18]   Ibid. Column 5, Line 56-57

[19]   "Google Acquires Glasses Patents," SEO by the Sea, April 13, 2012. Accessed March 26, 2014, http://www.seobythesea.com/2012/04/google-acquires-glasses-patents.

[20]   Poon, Aries. "Hon Hai Says It Sold Some Display Patents to Google" *The Wall Street Journal*, August 23, 2013. Accessed April 25, 2014 http://online.wsj.com/article/BT-CO-20130823-704061.html

Luk, Lorraine. "Foxconn Sells Communications Technology Patents to Google" *The Wall Street Journal*, April 25, 2014. Accessed April 25, 2014. http://online.wsj.com/news/articles/SB10001424052702304788404579523051086783712?mod=WSJ_hps_MIDDLE_Video_Top&mg=reno64-wsj

[21]   "Companies scramble for consumer data," *Financial Times*. Accessed June 13, 2013, http://www.ft.com/intl/cms/s/0/f0b6edc0-d342-11e2-b3ff-00144feab7de.html#axzz2W71Wg600. (Subscription required for access.)

[22]   Nick Pickels, "Google Glass: Orwellian surveillance with fluffier branding," *The Telegraph*, March 19, 2013. Accessed March 26, 2013, http://www.telegraph.co.uk/technology/google/9939933/Google-Glass-Orwellian-surveillance-with-fluffier-branding.html.

[23]   Alice Truong, "This Google Glass App Will Detect Your Emotions, Then Relay Them Back To Retailers," *FastCompany*, March 6, 2014. Accessed March 6, 2014, http://www.fastcompany.com/3027342/fast-feed/this-google-glass-app-will-detect-your-emotions-then-relay-them-back-to-retailers.

[24] John Zorabedian, "Spyware app turns the privacy tables on Google Glass wearers," nakedsecurity, March 25, 2014. Accessed March 25, 2014, http://nakedsecurity.sophos.com/2014/03/25/spyware-app-turns-the-privacy-tables-on-google-glass-wearers.

[25] Joseph Menn, "Analysis: The near impossible battle against hackers everywhere," *Reuters,* February 24, 2013. Accessed May 4, 2013, http://www.reuters.com/article/2013/02/24/us-cybersecurity-battle-idUSBRE91N03520130224.

[26] Jim Finkle, "360 million newly stolen credentials on black market: cybersecurity firm," *Reuters,* February 25, 2014. Accessed February 27, 2014, http://www.reuters.com/article/2014/02/25/us-cybercrime-databreach-idUSBREA1O20S20140225.

[27] "White hats to the rescue: Law-abiding hackers are helping businesses to fight off the bad guys," *The Economist,* February 20, 2014. Accessed February 27, 2014, http://www.economist.com/news/business/21596984-law-abiding-hackers-are-helping-businesses-fight-bad-guys-white-hats-rescue.

"Fighting China's hackers: Is it time to retaliate against cyber-thieves?" *The Economist,* May 23, 2013. Accessed February 27, 2014, http://www.economist.com/news/united-states/21578405-it-time-retaliate-against-cyber-thieves-fighting-chinas-hackers.

"Masters of the cyber-universe: China's state-sponsored hackers are ubiquitous—and totally unabashed," *The Economist,* April 4, 2013. Accessed February 27, 2014, http://www.economist.com/news/special-report/21574636-chinas-state-sponsored-hackers-are-ubiquitousand-totally-unabashed-masters.

[28] Eric Mack, "As Schmidt speaks of caution, Google Glass gets hacked," *CNET News,* April 26, 2013. Accessed April 28, 2013, http://news.cnet.com/8301-17938_105-57581724-1/as-schmidt-speaks-of-caution-google-glass-gets-hacked.

[29] Steve Henn, "Clever Hacks Give Google Glass Many Unintended Powers," *National Public Radio*, July 17, 2013. Accessed July 18, 2013, http://www.npr.org/blogs/alltechconsidered/2013/07/17/202725167/clever-hacks-give-google-glass-many-unintended-powers.

[30] "Saccade," *Wikipedia, The Free Encyclopedia.* Accessed March 14, 2013, http://en.wikipedia.org/wiki/Saccade.

[31] Susana Martinez-Conde, Jorge Otero-Millan, and Stephen L. Macknik, "The impact of microsaccades on vision: toward a unified theory of saccadic function," *Nature Reviews Neuroscience* 14 (2013): 83–96. doi:10.1038/nrn3405. Accessed April 9, 2014, http://www.nature.com/nrn/journal/v14/n2/full/nrn3405.html. (Payment required.)

[32] Jorge Otero-Millan et al., "Saccades and microsaccades during visual fixation, exploration, and search: Foundations for a common saccadic generator," *Journal of Vision.* Accessed May 27, 2013, http://www.journalofvision.org/content/8/14/21.long.

[33] Susana Martinez-Conde, Jorge Otero-Millan, and Stephen L. Macknik, "The impact of microsaccades on vision: toward a unified theory of saccadic function," *Nature Reviews Neuroscience* 14 (February 2013), 83–96, doi:10.1038/nrn3405 or http://www.nature.com/nrn/journal/v14/n2/full/nrn3405.html. Accessed April 17, 2014. (Payment required.)

[34] Ibid.

[35] Ann Marie Seward Barry, *Visual Intelligence, Perception, Image, and Manipulation in Visual Communication* (Albany, NY: State University of New York Press, 1997), 32.

[36] S.M.-C. and S.L.M., "Mystery Solved," *Scientific American Mind* 22, no. 5, (November/December 2011), 54. (Payment required.)

[37]   S.M.-C. and S.L.M., "Mystery Solved," *Scientific American Mind* 22, no. 5, (November/December 2011), 54. (Payment required.)

[38]   Susana Martinez-Conde, Stephen L. Macknik, Xoana G. Troncoso, and Thomas A. Dyar1, "Microsaccades Counteract Visual Fading during Fixation," *Neuron* 49, (January 19, 2006) 297–305, doi 10.1016/j.neuron.2005.11.033 http://www.neuralcorrelate.com/smc/files/publications/martinez-conde_et_al_neuron06.pdf. (Payment required.)

Susana Martinez-Conde and Stephen L. Macknik, "Shifting Focus," *Scientific American Mind* 22, no. 5, November/December (2011): 48–55. Accessed April 9, 2014. (Payment required.)

Susana Martinez-Conde, Jorge Otero-Millan, and Stephen L. Macknik. "The impact of microsaccades on vision: toward a unified theory of saccadic function," *Nature Reviews Neuroscience* 14 (February 2013): 83–96, doi:10.1038/nrn3405. Accessed April   9,   2014,   http://www.nature.com/nrn/journal/v14/n2/full/nrn3405.html. (Payment required.)

[39]   George Lakoff and Mark Johnson, *Philosophy in the Flesh: The Embodied Mind and Its Challenge to Western Thought* (New York, Basic Books, 1999) 13.

Gerald Zaltman, *How Consumers Think: Essential Insight Into the Mind of the Market* (Boston: Harvard Business School Press, 2003): 58.

Adam L. Penenberg, "A. K. Pradeep, Mind Reader," *Fast Company*, August 10, 2011. Accessed April 11, 2014, http://www.fastcompany.com/1772167/ak-pradeep-mind-reader.

[40]   Marisa Carrasco, "Visual attention: The past 25 years," Vision Research 51:13 (2011): 1484–1525. Accessed January 24, 2014, http://www.sciencedirect.com/science/article/pii/S0042698911001544.

[41]   Jack M. Wedam, US Patent 4,654,647, "Finger actuated electronic control apparatus" filed September 24, 1984 and issued on March 31, 1987. Accessed on May 20, 2014,
http://patft.uspto.gov/netacgi/nph-Parser?Sect1=PTO1&Sect2=HITOFF&d=PALL&p=
1&u=%2Fnetahtml%2FPTO%2Fsrchnum.htm&r=1&f=G&l=50&s1=4,654,647.PN.&
OS=PN/4,654,647&RS=PN/4,654,647

[42]   "Calibration," YouTube. Accessed May 8, 2013, http://www.youtube.com/watch?v=
xXmbgb00Kxo.

[43]   Hayes Solos Raffle, Adrian Wong, and Ryan Geiss, US Patent 8,235,529,
"Unlocking a screen with eye tracking information" filed November 30, 2011 and
issued on August 7, 2012. Accessed April 9, 2014, http://pdfpiw.uspto.gov/.piw?Page
Num=0&docid=08235529&IDKey=DB61E66D7179%0D%0A&HomeUrl=http%3A
%2F%2Fpatft.uspto.gov%2Fnetahtml%2FPTO%2Fpatimg.htm.

[44]   Raffle, Hayes Solos, Adrian Wong, and Ryan Geiss. US Patent 8,235,529,
"Unlocking a screen with eye tracking information" filed November 30, 2011 and
issued on August 7, 2012. Accessed April 9, 2014, http://pdfpiw.uspto.gov/.piw?Page
Num=0&docid=08235529&IDKey=DB61E66D7179%0D%0A&HomeUrl=http%3A
%2F%2Fpatft.uspto.gov%2Fnetahtml%2FPTO%2Fpatimg.htm

[45]   Raffle, Hayes Solos, Adrian Wong, and Ryan Geiss. US Patent 8,235,529,
"Unlocking a screen with eye tracking information" filed November 30, 2011 and
issued on August 7, 2012. Accessed April 9, 2014, http://pdfpiw.uspto.gov/.piw?Page
Num=0&docid=08235529&IDKey=DB61E66D7179%0D%0A&HomeUrl=http%3A
%2F%2Fpatft.uspto.gov%2Fnetahtml%2FPTO%2Fpatimg.htm

[46]   Ibid. Column 12, Lines 15-17

[47]   That is unless Google buries that technical information somewhere in the agreement
that few people would read and fewer would understand had they not read this book.

[48] This is a modification of a phrase, *"The Bible reads you,"* used by John Hagee, bestselling author of several books including *Four Blood Moons* (Brentwood, TN, Worthy Media, 2013).

[49] Zibreg, Christian. "Amazon smartphone head-tracking and 3D UI detailed, possible device prototype leaks" *idownloadlog*, April 15, 2014, Accessed April 24, 2014 http://www.idownloadblog.com/2014/04/15/amazon-smartphone-prototype-head-tracking/

[50] "General Information Concerning Patents, Specification [Description and Claims]," US Patent and Trademark Office, November 2011. Accessed January 28, 2014, http://www.uspto.gov/patents/resources/general_info_concerning_patents.jsp#heading-17.

[51] Hayes Solos Raffle, Adrian Wong, and Ryan Geiss, "Abstract: Unlocking a screen with eye tracking information," US Patent 8,235,529, filed November 30, 2011 and issued on August 7, 2012. Accessed April 9, 2014, http://pdfpiw.uspto.gov/.piw?Page Num=0&docid=08235529&IDKey=DB61E66D7179%0D%0A&HomeUrl=http%3A%2F%2Fpatft.uspto.gov%2Fnetahtml%2FPTO%2Fpatimg.htm.

[52] Hayes Solos Raffle, Adrian Wong, and Ryan Geiss, "Claim 8," Column 16 Line 35–39. US Patent 8,235,529, "Unlocking a screen with eye tracking information," filed November 30, 2011 and issued on August 7, 2012. Accessed April 9, 2014, http://pdfpiw.uspto.gov/.piw?PageNum=0&docid=08235529&IDKey=DB61E66D7179%0D%0A&HomeUrl=http%3A%2F%2Fpatft.uspto.gov%2Fnetahtml%2FPTO%2Fpatimg.htm.

[53] Hayes Solos Raffle, Adrian Wong, and Ryan Geiss, US Patent 8,235,529, "Unlocking a screen with eye tracking information," filed November 30, 2011 and issued on August 7, 2012. Accessed April 9, 2014, http://pdfpiw.uspto.gov/.piw?Page Num=0&docid=08235529&IDKey=DB61E66D7179%0D%0A&HomeUrl=http%3A%2F%2Fpatft.uspto.gov%2Fnetahtml%2FPTO%2Fpatimg.htm.

54   Gyroscope is briefly mentioned in column 4 lines 48–50, "a gyroscope coupled to the HMD may detect a head tilt, for example, and indicate that the wearer may be attempting to unlock the HMD screen." However, that is contrary to the title of the patent, which states that the screen is unlocked with eye-tracking information. There is no indication of how a unique head tilt, such as a slight tilt to the right and then the left, would unlock a screen. The premise of this patent was allegedly to make it easy to unlock a screen Just how that would be accomplished with a head tilt is not explained. This begs the question: why was it included in this patent unless it was to provide patent protection for Google Glass (but under a patent with a different title)? How GPS and accelerometers would contribute to unlocking the screen is not addressed in this patent.

55   V. S. Chinmay, "Sensors on Google Glass," The Code Artist Blog. Accessed January 26, 2014, http://thecodeartist.blogspot.com/2013/05/sensors-on-google-glass.html.

56   "Google Glass," *Wikipedia, The Free Encyclopedia*, http://en.wikipedia.org/wiki/Google_glass. Accessed January 26, 2014.

57   Ibid.

58   Hayes Raffle, Linkedin. Accessed April 20, 2014, http://www.linkedin.com/in/hayesraffle.

59   Adrian Wong, Linkedin. Accessed April 20, 2014, http://www.linkedin.com/in/almostsquare.

60   Ryan Geiss, geisswerks. Accessed April 20, 2014, http://www.geisswerks.com/ryan/pers.html.

61   "Consideration of Applicant's Rebuttal Arguments [R-9]," The US Patent and Trademark Office. Accessed February 27, 2014, http://www.uspto.gov/web/offices/pac/mpep/s2145.html.

[62] "Person having ordinary skill in the art," *Wikipedia, The Free Encyclopedia*. Accessed February 27, 2014, http://en.wikipedia.org/wiki/Person_having_ordinary_skill_in_the_art.

[63] "What the Eyes Reveal: 10 Messages My Pupils are Sending You," *PSYBLOG*. Accessed December 31, 2012, http://www.spring.org.uk/2011/12/what-the-eyes-reveal-10-messages-my-pupils-are-sending-you.php.

[64] Travis L. Seymour, Christopher A. Baker, and Joshua T. Gaunt, "Combining Blink, Pupil, and Response Time Measures in a Concealed Knowledge Test," Front Psychol. 2012; 3: 614. February 4, 2013. Accessed April 9, 2014, http://journal.frontiersin.org/Journal/10.3389/fpsyg.2012.00614/full.

[65] "Eye Heat Maps," Interactive Web Traffic. Accessed March 3, 2014, http://interactivewebtraffic.com/eye-heat-maps.

[66] Gus Lubin, Kim Bhasin, and Shlomo Sprung, "16 Heatmaps That Reveal Exactly Where People Look," *Business Insider*, May 21, 2012. Accessed March 4, 2014, http://www.businessinsider.com/eye-tracking-heatmaps-2012-5?op=1.

[67] "Milliseconds," *Wikipedia, The Free Encyclopedia*. Accessed March 4, 2014, http://en.wikipedia.org/wiki/Millisecond.

[68] Susana Martinez-Conde, Jorge Otero-Millan, and Stephen L. Macknik, "The impact of microsaccades on vision: toward a unified theory of saccadic function," Nature Reviews Neuroscience 14 (2013): 87. Accessed May 27, 2013, http://www.nature.com/nrn/journal/v14/n2/full/nrn3405.html.

[69] Susana Martinez-Conde, Jorge Otero-Millan, and Stephen L. Macknik, "The impact of microsaccades on vision: toward a unified theory of saccadic function,"

Nature Reviews Neuroscience 14 (2013): 86. Accessed May 27, 2013, http://www.nature.com/nrn/journal/v14/n2/full/nrn3405.html.

[70] "Covert," Oxford Dictionaries. Accessed May 27, 2013, http://oxforddictionaries.com/definition/english/covert.

[71] Susana Martinez-Conde, Jorge Otero-Millan, and Stephen L. Macknik, "The impact of microsaccades on vision: towards a unified theory of saccadic function," Nature Reviews Neuroscience 14 (2013): 85. Accessed May 27, 2013, http://www.nature.com/nrn/journal/v14/n2/full/nrn3405.html.

[72] Takemasa Yokoyama, Yasuki Noguchi, and Shinichi Kita, "Attentional shifts by gaze direction in voluntary orienting: evidence from a microsaccade study," Exp Brain Res 223 no. 2 (Sep 23, 2012): 291–300, doi: 10.1007/s00221-012-3260-z. Accessed May 27, 2013, http://www.ncbi.nlm.nih.gov/pmc/articles/PMC3475970.

[73] Jochen Laubrock, Ralf Engbert, Martin Rolfs, and Reinhold Kliegl, "Microsaccades Are an Index of Covert Attention: Commentary on Horowitz, Fine, Fencsik, Yurgenson, and Wolfe (2007)," Psychological Science. Accessed May 27, 2013, http://pss.sagepub.com/content/18/4/364.extract.

[74] Susana Martinez-Conde, Jorge Otero-Millan, and Stephen L. Macknik, "The impact of microsaccades on vision: towards a unified theory of saccadic function," Nature Reviews Neuroscience 14 (2013): 83. (revised June 3, 2002), doi:10.1038/nrn3405. Accessed May 27, 2013, http://www.nature.com/nrn/journal/v14/n2/full/nrn3405.html.

Ralf Engbert and Kliegl Reinhold, "Microsaccades uncover the orientation of covert attention," Vision Research 43, no. 9, (2003): 1035–45. Accessed April 9, 2014, http://www.sciencedirect.com/science/article/pii/S0042698903000841.

Ziad M. Hafed and Clark, James J. "Microsaccades as an overt measure of covert attention shifts," *Vision Research* 42 (2002): 2533–45. Accessed April 9, 2014, http://www.physiol-active-vision.uni-tuebingen.de/paper/hafed_vis_res_reprint2002.pdf.

[75] Susana Martinez-Conde and Stephen L Macknik, "Shifting Focus," *Scientific American Mind* 22, no. 5, November/December (2011): 48–55. Accessed April 9, 2014. (Payment required.)

[76] Jochen Laubrock, Ralf Engbert, Martin Rolfs, and Reinhold Kliegl, "Commentary: Microsaccades Are an Index of Covert Attention Commentary on Horowitz, Fine, Fencsik, Yurgenson, and Wolfe (2007)," *Psychological Science* 18, no. 4. Accessed April 17, 2014, http://lpp.psycho.univ-paris5.fr/pdf/PapersMR/2007/Laubrock-18-2007-364-6%20discussion%20367-8.pdf.

[77] Martin Rolfs, Ralf Engber, and Reinhold Kliegl, "Crossmodal coupling of oculomotor control and spatial attention in vision and audition," *Exp Brain Res* 166 (2005): 427, doi: 10.1007/s00221-005-2382-y. Accessed April 9, 2014, http://lpp.psycho.univ-paris5.fr/pdf/PapersMR/2005/Rolfs-166-2005-427-439.pdf.

[78] Ralf Engbert and Kliegl Reinhold, "Microsaccades uncover the orientation of covert attention," *Vision Research* 43, no. 9 (2003): 1035–45. Accessed April 9, 2014, http://www.sciencedirect.com/science/article/pii/S0042698903000841.

[79] Susana Martinez-Conde and Stephen L. Macknik, "Windows of the Mind," *Scientific American*, August 2007: 62. Accessed May 27, 2013, http://smc.neuralcorrelate.com/files/publications/martinez-conde_macknik_sciam07.pdf.

Z. M. Hafed and J. J. Clark, "Microsaccades as an overt measure of covert attention shifts," *Vision Res* 42, no. 22 (2003): 2533–45, doi: 10.1016/S0042-6989(03)00084-1. Accessed April 9, 2014, http://www.sciencedirect.com/science/article/pii/S0042698903000841.

Ziad M. Hafed, "Alteration of Visual Perception prior to Microsaccades," *Neuron* 77 (2013): 775–786. Accessed May 27, 2013, http://www.cnbc.cmu.edu/braingroup/papers/hafed_2013.pdf.

[80]   Susana Martinez-Conde and Stephen L. Macknik, "Shifting Focus," *Scientific American Mind* 22, no. 5, November/December (2011), 48–55. Accessed April 9, 2014, http://www.scientificamerican.com/magazine/mind/2011/11-01.

[81]   Mark Prigg, "Glass without the glasses: Google patents smart contact lens system with a camera built in," *Daily Mail*, April 14, 2014. Accessed April 14, 2014, http://www.dailymail.co.uk/sciencetech/article-2604543/Glass-without-glasses-Google-patents-smart-contact-lens-CAMERA-built-in.html.

[82]   Brad Stone, "Inside Google's Secret Lab," *BusinessWeek*, May 22, 2013. Accessed April 18, 2014, http://www.businessweek.com/articles/2013-05-22/inside-googles-secret-lab.

www.ingramcontent.com/pod-product-compliance
Lightning Source LLC
Chambersburg PA
CBHW071820170526
45167CB00003B/1383